I0473449

Fire Stick

2019 User and App Guide: Get the Most out of your Amazon Fire Stick! With Step-by-Step Instructions

by

John Sakuma

ISBN: 9781071168196

Imprint: Independently published

Introduction

Congratulations and thank you for purchasing the *Amazon Fire Stick Guide.* It's my firm belief that it will provide you with all the answers to your questions.

Since its introduction, the Amazon firestick has significantly improved. Amazon has worked to offer the best customer experience to Firestick owners. With technology changing every day, there has been a need for improvement in the Firestick since its inception.

There is a variety of content to choose from with Firestick. There are news channels, music channels, movies, online radio, and many more. Channels include WatchESPN, Prime Video PBS, Hulu Plus, A&E, Plex, Spotify, Prime Music, PBS Kids, Vevo, ShowTime, AnyTime, Twitch, NBA Time, Disney Channel and Netflix.

CHAPTER 1

Amazon Fire TV Stick Explained

What is it?

Amazon Fire Stick is not just a tedious addition to TV. This fundamentally new device has expanded the range of streaming technologies for home entertainment.

Fire Stick connects to your TV's HDMI port to give you access to a world of favorite games, music, movies, photos, subscription services and TV shows.

Imagine going with a TV to any corner of the globe. Is this an attractive offer? This is exactly the bonus you get when you buy a Fire TV Stick. It is so small that it can fit in a handbag. Fire TV Stick contains all the content and is fully portable. Just unplug the Firestick and carry it around with you. Connect it to any compatible TV to get immediate access to the content that interests you.

Amazon Prime customers get special benefits. If you have a registered account, you immediately get access to all the content found exclusively on Amazon Prime. All this happens immediately when you connect the Fire Stick to a HDMI port.

The Amazon Fire TV Stick Revolution

The Fire TV Stick 1st Gen

The 1st generation Fire TV Stick, codenamed "Inigo," included a remote but had no voice search. Later, on the 1st of November in the year 2014, Amazon introduced a version of the Fire TV Stick that was smaller and codenamed, "Montoya." The device had a HDMI port plug-in device that replaced almost all of the bigger Fire TV's functionality. Although the hardware is a little bit different, it came with a 1GB of RAM, internal storage of 8GB, weight of 25.1 g (0.9 oz.) and a dual-core processor. It also had Bluetooth 3.0 and dual-band wireless.

The Fire TV Stick 2nd Gen

The 2nd gen Fire TV Stick came with Alexa voice remote and "Tank" was its codename. Tank came with new updates like a Quadcore processor and Bluetooth 4.1 and Wi-Fi hardware. The storage remained the same and its weight was 32.0g (1.1 oz.).

The 3rd Gen Fire TV Stick

On October 31, 2018, Amazon stated its intension to offer the Fire TV Stick 4K with all-new Alexa voice remote. Included for the first time were volume, power and mute buttons. The launch date was October 31, 2018.

This version came with an HDR and Ultra HD streaming via HDMI dongle. It is compatible with Dolby Vision, HDR10+ and Dolby Atmos. This provides an exceptional audio and video experience. It syncs with Alexa remote with facets such as power, muting buttons, infrared and Bluetooth. Moreover, the in-app Alexa control will be added to expert video services such as Sony Crackle, HBO Now, AMC and many more.

It has a 1.7GHz processor which is faster than that of the 2017 model. This version has a smoother response and loads quicker than its predecessor. Germany, India and the UK released the Fire TV Stick 4K on November 14th, and Japan got it by the end of 2018.

What's in the Box

The box usually has the following items:

- Amazon Fire TV Stick
- Power adapter for Fire Stick
- Alexa Voice remote control
- An extender cable
- Two AAA Amazon Basics batteries

Fire Stick Features

The Fire Stick has some pretty basic features:

- A gig of memory
- Voice support with the Alexa remote
- VideoCore4 GPU
- Dual-band, dual-antenna Wi-Fi connection
- 8 GB of internal storage

This is typically what you get, and if you want to upgrade, it will cost a little bit more. You'll be able to create and enjoy a streaming service and adventure, allow a wonderful and worthwhile endeavor.

Other features have been added as well. The new Fire Stick now offers volume, mute and even power buttons on Alexa remotes. The Fire TV Stick is now in 4K and supports all the major HDR standards, including HDR10, HDR10+ and even Dolby vision and Dolby Atmos audio. This is a new feature that provides dynamic metadata, complete with details about the brightest and darkest aspects of an image and whether it is compatible with the TB. It's a good way to determine whether you're fitting the standards you now have. Supporting content for this standard is important, and these features have changed the game.

It also allows you to game the way you want. Some of the titles don't work with the gaming remote, but there are a few that actually work quite well, such as Minecraft, the Bard's Tale and Star Wars: Knights of the Old Republic. It helps you really get your gaming on and assists in choosing the best activity for you.

If you are really not into using the Alexa remote that comes with it, there is a side feature that connects the phone app to an iPhone or even an Android as a replacement for it. It's a good alternative if there is any chance that it will end up somewhere else. Getting the app is a good option for those who don't want to spend time struggling to find a remote.

Voice control is the name of the game, and, essentially, if you've got Alexa programmed, you can control the volume, channel changing or whatever. If you want a super easy means of controling everything in your home, the Fire Stick is the way to go. Plus, with the integration of Alexa, it allows you to create a hands-free life.

When it comes to bringing forth good ideas for your Alexa device, the Fire Stick is here. There are so many unique and amazing features to choose from. Here we highlight just a few that

will ultimately change the game for you as a
user.

Comparison of Fire TV Stick to Rival Devices

Roku and Chromecast are its rivals, but with this device, you just plug it in and it's even available with a handy Bluetooth remote. Once plugged in, you're brought to the setup process. It's essentially a repository for everything you have; and once you're in there, you can access thousands of games and apps. With just the press of a button, you can open up whatever you want. You have the advantage of choosing what you want instead of having to pay for costly packages.

The remote is huge advantage. If you don't like excessive menus and would rather groan than work with one, then the Alexa remote is going to be your best friend. The best thing about it is that by pressing a button and saying what you want, you can let it search and it gives you the full advantage of choosing what you want. You'll soon realize that it's super simple, and you only have to navigate a few menus and not a ton of buttons.

So, yes, you can bring it with you and log into Wi-Fi, whether at a hotel, a friend's house or wherever. You'll then be able to have all of your favorite shows on the go. You can also add Kodi. We'll discuss that later on.

How it Works

First of all, you enter your Amazon account information and connect the remote. There are six main sections: search, home page, movies, TV shows, applications and settings. Users always start from the home section. To navigate through the various sections and content, you use the navigation pad on the remote control. The trackpad part in the middle acts as a selection button.

Instead of using different buttons on the console itself, you can perform numerous tasks by pressing the microphone button and talking to your voice assistant. You can command it with your voice to play a TV show or movie, search for content, search applications, control playback, launch applications, control live applications, and more.

Fire Stick users have a great opportunity to use their voice assistants to search for content. You can ask it to solve complex problems. For example, you can select playlists or specify music of a particular genre in applications. But that's not all: your voice assistant is positioned to be a digital assistant with even more features.

Using your voice assistant, you can manage meetings on Google or Microsoft calendars, manage smart home devices, or perform search queries. If you connect additional Skills, you can order food through Lieferando or a taxi through Mytaxi. Speech recognition works fine, but Alexa's response to voice calls is a different matter. The user should know exactly which Skills are installed in your voice assistant. Otherwise, voice control may reach a deadlock. Your voice assistant on the Fire TV Stick has less functionality than on the Echo. For example, you cannot adjust the volume on the Fire TV Stick, and the alarm clock and timer are not supported. There are quite a few such "pitfalls."

On the other hand, the possibilities of Alexa on the Fire TV Stick are greater than the possibilities on the Echo. The fact is, Alexa can respond to a command not only from a voice but also visually. The weather forecast is not only pronounced but is also displayed on the screen. Amazon has redone the temperature output from Fahrenheit to Celsius.

Your voice assistant can recall, for example, the results of Bundesliga football matches. But other sports are not yet supported. The range of functions is very limited. However, if you stick to voice commands that your voice assistant understands, the system will work reliably. As for navigation, it is easier and quicker to get to the desired point using the remote, rather than voice. Finding a movie or music with the help of your voice assistant on Fire TV Stick is also easy.

CHAPTER 2

Fire Stick and Fire TV Compared

You have probably looked online and have seen sites using both the terms, Fire Stick and Fire

TV. But what is the difference? Which one will fit you better? Well, read on to find out. Here, we're going to explore both and determine the best option for you as a user.

The Fundamentals

So, you have Fire TV. This is a media player, and it is part of the Fire TV ecosystem. It has so many streaming benefits, including high-tech features and the ability to deliver a ton of entertainment and experience, including 4K ultra HD streaming. Plus, you have over half a million films and TV shows. Furthermore, it has the Alexa integration you'll love. It's super small, and comes with standard HDMI ports and a microUSB port for side loading apps. There is also an Ethernet adapter to help improve the seaming qualities; and since it's always behind your TV, there's no reason to worry about others seeing it.

It's very easy; all you need is plug it into the HDMI port and then you're connected. It works with Alexa, and the voice remote period will help you control the content you want to create seamless streaming. It can be paired with echo devices as well, so you can use it for full control, and it has a pretty easy setup.

Now let's evaluate the Fire Stick. Essentially, it is similar to the Fire TV, but for half the price. While the Fire TV is close to 100 bucks, this one is 40 for the Ultra HD option and 10 dollars cheaper for the first edition. The cheaper option also offers the same number of movies and TV shows, all on a little stick. It doesn't take much to put together or operate, allowing you to have the best steaming capabilities possible. It's a miracle what it can do, with so many cool benefits.

So, which is better for you? Well, it depends on what you want and the power you desire. It's debatable as to the better option, so we'll explore a couple of the ways one might benefit you over the other.

The Fire Cube

So, you have Fire TV and the Fire Stick, but when you look online, chances are you see the Fire TV cube and wonder what the heck it is. Well, fear not, for we will tell you whether it's worth everything it claims to be.

The Fire Cube is basically the love child of a Fire TV and an Echo and Echo Dot. It allows you to stream movies and TV from different services, and you can control the information in your home, whether it be your smart home devices or just having Alexa control what is playing on your Fire TV.

How it works is simple. The Fire TV Cube is made to hear you even if there are other entertainment options in this room. Essentially, once it hears the word, Alexa, it will pause that content, mute it and then listen to your commands. It's quite nifty and pretty simple to utilize.

It has some cool touches as well. If you are continuing a show that has started, it will pick up right where you left off. If you are searching for movies and shows on streaming services, the Fire Cube will essentially list every single option and choose where to watch them. For example, if there is a show on Netflix, but it costs extra compared to watching it on Prime, the Fire TV Cube will choose the right device.

Meanwhile, the device is getting a whole new set of updates. For example, if you are tired of saying, Alexa, all the time, you can essentially utilize the follow-up mode that has Alexa continue to listen for further commands or requests. Alexa will do a secondary command, saving you from having to shout Alexa all the time.

Alexa announcements are compatible with the Fire TV Cube and the Fire Stick as well, allowing you to cast a message to the other Alexa devices like an intercom. It literally does Echo show dropins without needing an Echo show. So, if you want to tell people that the movie is about to start, Alexa will send out that message to every device in the home. You can just have Alexa announce or broadcast it, and the Fire TV Cube will receive the announcement from smart home devices. These announcements are good, especially if you have a camera; the Fire TV Cube will actually let you know if something is there. Alexa essentially takes the Fire TV Cube and helps you create a more integrative system for all your devices.

In short, it has a few neat touches, and just like with the Echo, the Fire TV Cube can be a smart home hub, allowing you to control different options. Unlike Fire TV and the Fire Stick that don't directly control a smart home, it actually allows you to have complete control of your devices. If you are looking to get an Echo or want the Alexa touch without needing to pay a hefty price for it, this is your best bet.

Processing Power

Processing power enhances the experience, and it's important that you consider it if you're going to be a serious streamer. The Fire Stick has an ARM cortex A-7 that clocks out at 1.3 GHz, while the fire cortex goes a little bit higher. It's only a. .2 difference, but definitely a bit of a jump. The Fire Stick has a lower processor, about 32 bit rather than 64, which is more of a decisive factor. You will notice the performance differences, and because the Fire TV has more RAM, it plays a good role in regard.

Entertainment Value of the Fire TV

The Fire TV has a bit of extra value in giving you the streaming benefit of feeling like you are at the theater; and with live TV, it is definitely pretty nice. Plus, with Fire TV, you can get 4K ultra HD models, and they definitely give you a more advanced picture quality than, say, watching on your tablet or smartphone. It works to 60 FPS too, so you won't have to worry about the screen getting all fuzzy and any buffering. It also works with Dolby systems, including Atmos, Digital and Digital plus; and you'll be able to create the perfect entertainment experience for you and your home.

Storage

Storage is the one area on Fire TV devices that does not really change. They both have 8 GB of internal storage, and there's no external storage possible. The Fire Stick will save you some money, though, and they weren't created for storage purposes. That's because you can side load virtually any app, and you'll need it for online streaming, making 8GB perfect. But maybe they'll create an update where more storage is possible in the future. It certainly wouldn't be that much of an issue, but it would be a nice little addition.

Size

While both of devices tend to operate on smaller systems, there is the extra benefit of the Fire Stick being a bit smaller. Fire TV is a dongle, but the Fire Stick is literally a stick. The TV is obviously a little bigger than the stick, and that makes it a lot more compatible. But both fit pretty well in bags. The Fire TV should be kept in a bag, whereas the Fire Stick is so small that it can fit in your pocket. It's not really a factor, because both work in the same way. Thankfully, the Fire TV isn't a box, so you can leave it behind if you really want to. It's up to you.

However, if you are looking for the ultimate in portability, I suggest the Fire Stick. It is pretty good if you need something you can take easily and put away when you're done

Both Support 4K and HDR nowadays

The earlier models not so much, but Fire TV and the Fire TV stick both support 4K resolutions and the HDR protocols you want. The Fire TV stick does have less resolution on the standard model, only running up to 1080p at the max, and it doesn't handle the 4K UHD. But, the new models do, so both are pretty good if you want something strong.

However, they have another differentiation. The Fire TV stick and Fire TV both support the 4K resolution needed, along with HDR10, allowing for richer color. But there are two other major HDR means, HDR10 and Dolby vision, and only one the Fire TV supports is the latter. This is a big thing if you're super into videos, but it's not that much of a distinction at present. However, there is a chance for a firmware update, but from the quality perspective, the Fire TV stick is slightly higher in terms of firmware and other options.

Alexa

Is Alexa any different from either of the devices in question? Well, if you have Alexa, you can actually check the weather, manage lists, look up scores, research the traffic, and even create different playlists on your TV. You can do this on any Fire TV model, using Alexa. The Fire Stick has some Alexa integration, but not as much. The newest Fire TV handles this a little bit better since the Cube is actually a speaker. This means you can talk to it and give instructions without needing to hold down the button for Alexa on the remote. It also has better integration with smart home gadgets, so the Cube is the better option if you want to control the lights, temperature and several devices. If you want an Alexa speaker and have a streaming device, then get the TV Cube; but if you're already got an Echo, don't need to speak to it and are fine without smart integration, the Fire Stick is the better choice.

Operating System

There is a distinctly different OS between the Fire TV and the stick. The TV runs on Fire OS6, what you see in the Nougat OS on tablets and smartphones; but the Fire Stick runs at the older Fire OS5 version. The difference is simple: the Os6 fixes some patches and bugs, and the average person probably won't see them. You may not even see any distinctions one over the other one. They don't have remarkable distinctions, so you're not getting anything super different; there might be some changes in the future, but for now, they're one and the same.

The Fire Stick and Fire TV are similar in nature, and you'll be surprised at how little difference there is. If you're looking for the right system for you, either of them will suffice; you'll get benefits from both.

CHAPTER 3

Setting up the Amazon Fire TV Stick

Viewing of TV is swiftly changing everywhere in the world with the advent of video streaming services such as Amazon Prime, Hotstar, Netflix, Hulu, etc. DTH may very quickly become a thing of the past, and it will be for the better. Other than the provision of better and greater video streaming services, viewers have a true choice. They can pick anything they desire from hundreds of thousands of documentaries, movies, TV shows, etc.

Amazon Fire Stick is an extraordinary product, providing a multitude of offerings like Amazon Prime Video, Hotstar, Netflix and Gaana, among several others. But this isn't the sole highlight of the Fire Stick. The Fire Stick remote control now comes with Alexa support. So, you can just sit comfortably on your couch and control everything via voice command.

If you're going to buy the Amazon Fire TV Stick or have already bought it and are in a quandary about setting it up, then stay calm and relax. We're right here to walk you through your first setup of your new Amazon Fire TV Stick as a beginner on your TV. Just observe these steps carefully and the whole ordeal will be a walkover.

First of all, before you begin your setup, you are require to get things ready.

Note that the following are not contained in the Fire Stick Box:

A compatible TV: this is the primary requirement. You need at least an HDTV that uses an HDMI port in order to use your Fire Stick.

An Internet connection: The Fire Stick doesn't come with an internet connection. You will have to connect it to a live Wi-Fi network strong to enable it to stream HD video appropriately.

An Amazon account: you will have to sign up an Amazon Fire TV to an Amazon account. If you ordered your Fire TV through your Amazon account, it will already be registered. Note that you can change it anytime you so desire.

At this point, you are set and good to go. You can now begin your Amazon Fire Stick setup. For beginners, the setups process has been divided into four related phases to make it easy.

Phase 1: Getting Started

1. Bring up your Fire TV Stick.
2. You will see a port for micro-USB with an HDMI.

3. Plug your USB power cord into the port of the Fire Stick and the opposite end to your Power Adapter.

4. Plug in the Adapter to the power socket. You are advised to make use of a Power Adapter and not the normal powered USB Ports provided by some TVs.

5. You can now plug your Fire Stick into your TV's HDMI port.

6. Put your TV on and select HDMI input Channel.

7. The Fire TV Stick logo will appear on the loading screen. It will take a little while to load for the first time, so don't panic.

8. At this point, you will be required to set up the remote control of your Fire Stick.

9. First, you have to insert the included 2 AAA batteries in the box. But, if your remote fails to pair automatically, you must tap and hold the Home button on your remote for at least 10 seconds. This will immediately transfer your Fire Stick to Discovery Mode and pairing will exist.

10. The moment you have paired your remote with your Fire Stick, press the button "Play/Pause" to begin the setup process.

Your Fire Stick will now prompt you to select your desire language. To do this, you will use the navigation key on your remote to go directly to your desired language. Highlight it and click the Select/OK button on your remote to set this language.

Tips: your Select/OK is the circular button found on your remote, walled inside the navigation buttons.

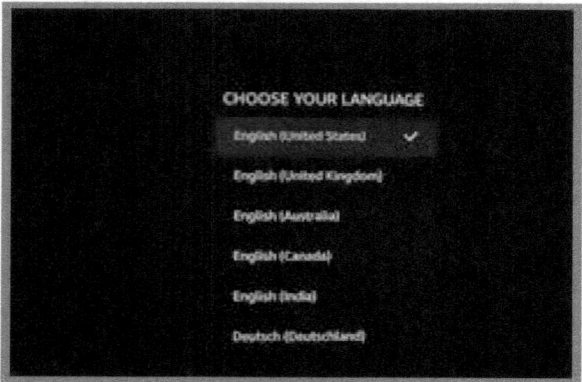

Phase 2: Connect your Amazon Fire Stick to a Wi-Fi network

1. At this point, your Fire Stick will now scan for accessible Wi-Fi networks that exist within range.
2. On the list of networks obtained, select your Wi-Fi and enter your SSID and password if asked.

Phase 3: Sign up and register your Fire TV/Stick with your Amazon Account

1. Now you are required to register your device to your Amazon account.

2. If you have ordered your Amazon Fire Stick from Amazon, your device will already be registered. If it is not a big deal or you desire to register via another Amazon account, you can simply follow the on-screen directives to register and de-register your Fire TV.

If your Fire Stick is not registered already, you will come across the window below. But if you are already signed up for an Amazon account, choose the first option. Otherwise, create an Amazon account just by pressing the option "I am new to Amazon."

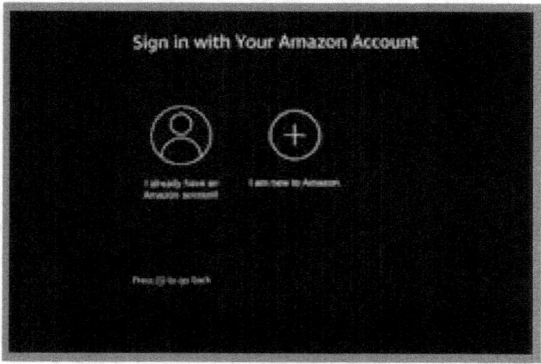

The Fire Stick will now be registered. Just wait a few seconds.

Phase 4: Winding Up your Setup Process

1. When your registration is done, you will be asked by your Fire Stick whether you would like to save your password for Wi-Fi in your Amazon account.

2. If you use more than one Amazon device and register them to the same Amazon account as your Fire Stick, it would be wise to select (Yes) to let your devices connect faster to your Wi-Fi network. But in the case where you have only one Amazon device and will never desire to purchase any in the future, just select (No).

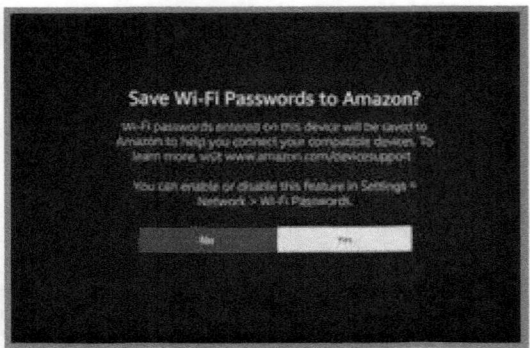

3. If you desire to activate Parental Controls on your Fire Stick, press Enable Parental Controls on the next screen.
4. Next, you will be prompted to setup your PIN.
5. You will be requested to enter your PIN for every activity on your Fire Stick.

If you don't want Parental Controls on your Fire Stick, just press "No Parental Controls" to continue.

There you have it: you are finished. While your Fie Stick is runn ing in the background, you will be offered a brief video clip, displaying tips on navigating and using your Fire TV.

Configuring Your Device

Although most settings on your Fire Stick are automatic, there may come a time when you would like to make custom changes. All this functionality is handled through settings. There are a variety of areas that can be changed within the settings:

- Display & sound
- Parental Controls – (Free Time is only available on Amazon Fire TV)
- Controllers & other Bluetooth devices

- Applications
- System
- Help
- My Account

These areas will be further discussed later in this chapter. They are important because they essentially govern your Fire Stick experience. Although this chapter covers the basics, the information can still be useful in customizing your experience.

Display & Sounds

This area of the device allows the user to perform a multitude of functions:

- Set a screen saver
- Configure the display
- Set up device mirroring with a compatible device
- Manage audio settings
- Enable discovery of Amazon Instant Video playback
- Obtain photos on the Amazon Fire TV device from nearby compatible mobile devices

Parental Controls

Parental controls make it easy to let your child into the world of digital media. The parental setting on the Amazon Fire Stick gives you the ability to not only block purchases your child may make, but it can also restrict access to a number of games, movies, television shows, apps, photos, and so much more. It is important to keep in mind that 3rd party applications cannot be blocked.

The parental controls are locked within the application using a PIN. This PIN will not work when attempting to enter it with a 3rd party remote – only online or physical Amazon remotes can be used. To set up your parental configuration, follow the steps below.

1. From your HOME screen, navigate to Settings.
2. From the Settings menu, select Parental Controls.
3. Enter the PIN for Parental Controls. This is the same PIN used for items like Amazon Instant Video.
4. Utilize your remote to select On or Off for parental controls.
5. After your PIN has been completely set up, you are able to choose the options for which the PIN is needed:
    i.   Mandating that all purchases require a PIN
    ii. Mandating that a PIN is needed for Amazon Instant Video only
    iii. Blocking the ability to purchase or even view certain content types, such as games, apps or photos
    iv. Changing your PIN number

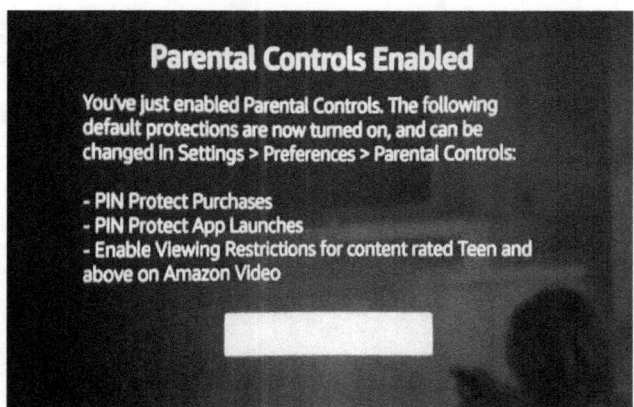

There are a variety of commands to be used in parental control. These include the following, taken from Amazon.com:

- PIN Protect Purchases – PIN protected purchases ultimately require that before a purchase is completed, you must verify your identity using the PIN previously selected.
- PIN Protect App Launches – This functionality requires that before opening a certain type of application, which include pre-installed applications, you must use your PIN. You still have the ability to browse the applications. However, they will not open without the entry of your PIN.

- PIN Protect Prime Photos App – this function needs the PIN to open the Prime Photos app. This app lets you view personal videos and photos.
- Viewing Restrictions – This allows you to block various TV Shows and movies from stock Amazon videos or other 3rd party apps, depending on their category of rating.
- Change your PIN – this lets you change the PIN of your Amazon Video on your Fire TV.

CHAPTER 4

Using the remote control

Your Amazon Fire Stick remote has many functions for each of its buttons. The table below shows the function of each buttons on your remote. (This list is taken from Amazon.)

Button	Function
Voice *Only available on Alexa Voice Remote versio ns* 🎤	The microphone is built-in, allowing you to interact with Alexa through your voice
Directional Navigation ◉	These buttons allow you to scroll left, right, down and up
Select ◉	The button lets you select the highlighted item on the screen
Home ⌂	This button takes you to the Home screen
Menu ☰	This button allows you to have increased options for the function or item currently highlighted

Back ⊃	This button takes you to the previous action or screen
Rewind ⊲⊲ Pause/Play ▸॥ Fast Forward ▸▸	The media control buttons allow you to fast-forward, pause, rewind and play an audio or video For video playback, press the Forward or Rewind butt on. Pressing once skips 10 seconds forward or back.

Source: amazon.com

Navigating Your Fire TV Device

If you have recently purchased your Fire Stick, you will notice that this remote allows speaking commands to your Fire Stick. The following list includes the basic activities the Fire television can perform and how to perform them.

Function	Instructions
Navigating to the *HOME* screen	To get back to the HOME screen from any screen on the Fire TV, simply press the ◎ on the remote.

Returning to the previous screen position	To return to a previous screen, simply select the back arrow. ⤶
Selecting an item to watch – a movie, television show or even a standup comic	Once you have landed on an item of interest, simply highlight the item using your remote's arrows to get to and push the middle circle on your remote.
How to bring your Fire TV out of sleep mode	To awaken your Fire TV device, push any button on you remote.
Putting your Fire TV into rest/sleep mode	To put your Fire Stick into sleep mode, you can simply leave the device inactive for thirty minutes. You can also manually put your device into sleep mode by navigating to *Settings -> System -> Sleep*
Find content previously purchased through Amazon	Your purchased content can be found in a variety of locations based on the type of media. The library breakdown is as follows:

	· Video Library – includes items that have been purchased pr rented. This library can include movies and TV shows. Items found within streaming apps like Hulu or Netflix are not available from the video library · Games – includes any games purchased on your device. · Apps – includes any applications purchased on your device · Music – includes any music purchased, imported or held via Amazon Prime.
Customizing & removing Home screen content	To remove items from your recent list, which appear on your HOME screen, scroll to the designated item and select "Remove from Recent."

	Recommendations that appear on your home screen can be removed as well by simply selecting "not interested" on the item.
Changing the Screen Saver settings	Your screensaver can be changed by following the instructions below: From the Home screen -> Settings -> System ->Screen Saver Here you can update the slide style, speed and start time of the Screen Saver. You can also select which Photo Album to use.
Turning off your Amazon Fire TV	Turning off your TV is easy. The remote now comes with a standard power button instead of the untraditional method of removing the power adapter from the Fire Stick Device.

	An alternative to turning off your device is placing it in sleep mode, which can be done from the settings.

Pair with Your Fire TV Remote App Along with Your Fire TV Stick

By the time we begin using the Fire TV Stick, there's one more you need to do. You must download and pair the Fire TV Remote app.

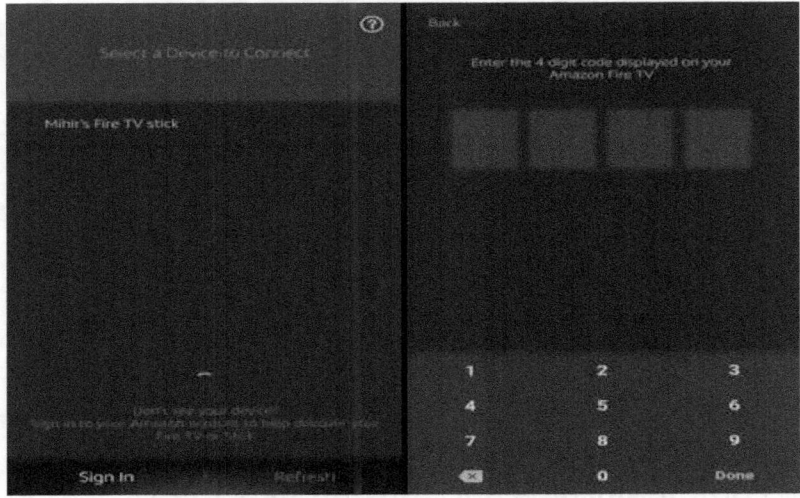

As good as the Alexa Remote is, the setup procedure must display how annoyingly hard it is for typing. A better solution is Amazon's Fire TV Remote app, also profitable for Android and IOS. So, download and then install it on your smartphone.

- First, download the app, Amazon Fire TV Remote, for iPhone/iPad or for Android.
- Connect your phone to the same network. Open the app and connect it.
- On the phone, type the 4-digit code that appears on your TV screen. This should pair the two devices successfully.

Voice control and Alexa

You had the ability to control the first generation of Fire Sticks with a remote control which took a lot of time and was not always convenient. This changed with the second-generation Fire Stick, which now comes with a voice control remote. Not only does it allow you to enter information easily, but now you can also use the personal assistant, Alexa. Thus, it opens completely new horizons for users.

Unlike other Alexa-compatible devices, like the Echo, the Fire Stick remote doesn't use a wake word. To activate a voice control session, you will have to press and hold the microphone button (over the directional pad) and speak into the remote.

Users of a remote app on a smart phone not only use voice commands but can also type in names and titles using the on-screen keyboard, something that makes it even more appealing than a voice-controlled remote. This app is available for both Android and iOS devices and hears you through your phone's speaker.

If this is your first attempt at Alexa or you have never had an Alexa device, don't despair; this chapter will aid you in knowing just what Alexa can do and how it can work for you. Alexa allows you to do a multitude of things including shopping, finding music and finding television shows, and so much more.

The Fire Stick utilizes a wide variety of skills. These skills may include functionalities paired with stores like Amazon, Uber, Dominoes or even StubHub. These skills aid the user in adding additional functionality to Alexa. These items are added via the Alexa App for your mobile.

Alexa lets users interact not just with music, but also with books, podcasts, radio programs and much more. A premium subscription may be needed to access certain content. If a subscription is not held, a personal music library can be connected using Alexa.

Subscription services compatible with Alexa include the following:

- Amazon Music Unlimited
- Spotify Premium
- Prime Music
- Audible
- iHeartRadio
- TuneIn
- Amazon Music

Audible and Kindle are also applications that can be used by Alexa to stream audiobooks, and newspaper and magazine subscriptions.

Playing music is a simple task for television. Sample commands are given below:

"Play Trey Songz Radio on Pandora"

"Play Spotify" - This instruction simply plays music where you last left off within the application, Spotify.

The below commands play in whatever your default application is:

"Alexa, play Taylor Swift"

"Alexa, shuffle Trina. "

Alexa also provides regular news, weather and sports updates. Alexa can give answers to simple questions such as holiday dates, information involving television or movie programs and on well-known actors or actresses.

Keep on schedule and on time with Alexa. The user has the ability to add and review any calendar events through Alexa's linked calendar.

Alexa can be used to add or review events. To add events, you must link the calendar within the Alexa mobile application. To do this, follow the following procedure:

Settings> Calendar > select Google Calendar. Choose "Link Google Calendar Account" to link your active Google calendar.

Try these commands when interacting with the calendar.

"When is my next event?"

"What's on my calendar on [day]?"

Alarm settings can also be used within the Alexa application:

"Wake me up at 7AM."

"What time is my alarm set for today?"

"Set a repeating alarm for Mondays at 5AM."

Aside from calendars and alarms, lists can also be managed using Alexa. These may include shopping or to-do lists. There is also the option to link third party list services.

CHAPTER 5

Accelerating Fire Tv Stick's Settings

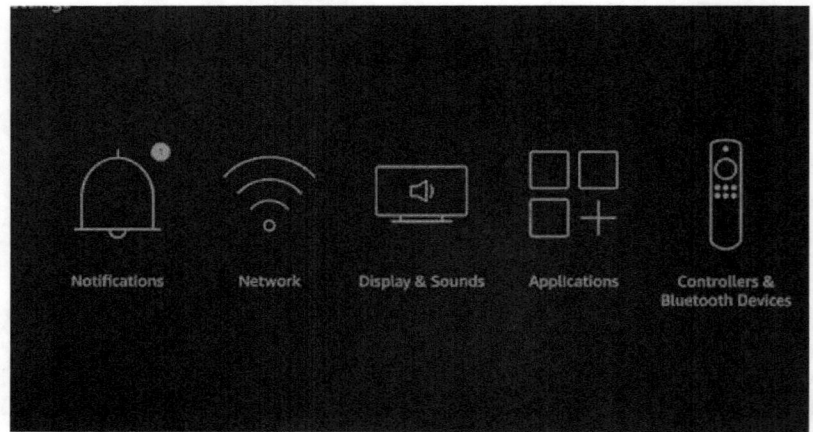

On the main menu, the last alternative allows you to take charge of a number of Fire Stick functions:

- Altering DNS servers and connecting to hidden networks
- Changing the setting of the screensaver
- Changing video and audio settings
- Uninstalling and managing apps
- Changing the nickname of the Game Circle
- Changing the language and time zone
- Changing the Prime Photos app's setting

Mirroring the screen to Fire TV Stick

Fire Stick allows you to mirror your tablet or phone screen so that what you see on your tablet or phone will be displayed on the TV screen. This is possible via Mira cast, which is supported by most tablets and smartphones.

1. Get to Settings, then Display, and enable Mirroring
2. Get to the phone's Setting, select Display and then Wireless Display
3. Make sure you have enabled it and among the options available, select Fire Stick

Disabling in-App Purchases

If not keenly checked, the in-app purchases of the Fire Stick can accumulate quite a bill. It is advisable to disable the in-app purchases if the Fire Stick will be used by children.

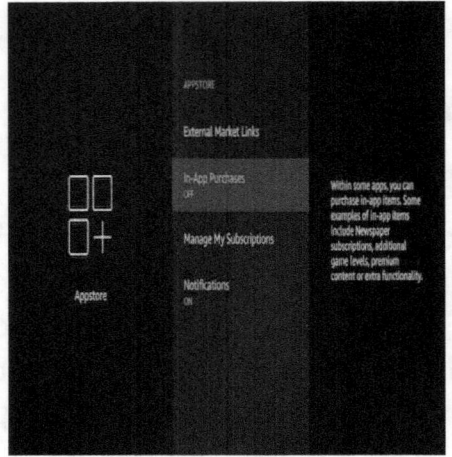

1. Get to Settings, select Applications, select Appstore and choose In-App Purchases

2. Switch it off

Data Monitoring / Enabling and Disabling

Parental Controls

In the setup, you will have the alternative to skip or activate data monitoring and parental controls. Turn the switch just in case you think otherwise.

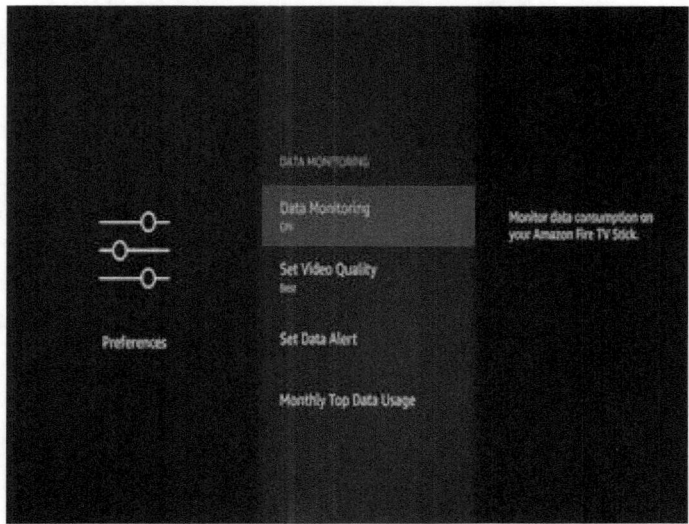

1. Move to settings, select Preferences and then select Parental controls

2. Switch it off or on.

Enabling the Fire TV Stick Accessibility

Alternative

The Fire TV Stick has some features for people with sound impairment. These features make it user-friendly for them and include high-level contrast text, VoiceView, and closed caption subtitles.

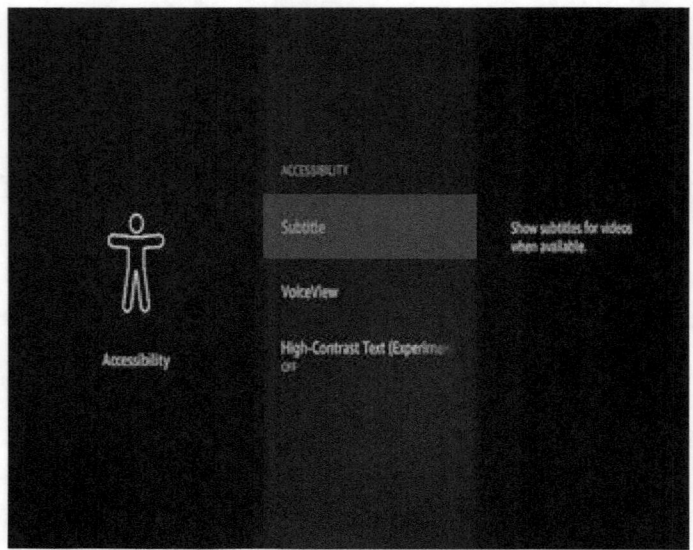

These alternatives can be enabled by: Settings > Accessibility.

Resetting Fire Stick to Factory Settings

You do not have to worry if you mess up anything, because you can easily restore the Fire Stick to factory settings to be just the way you bought it.

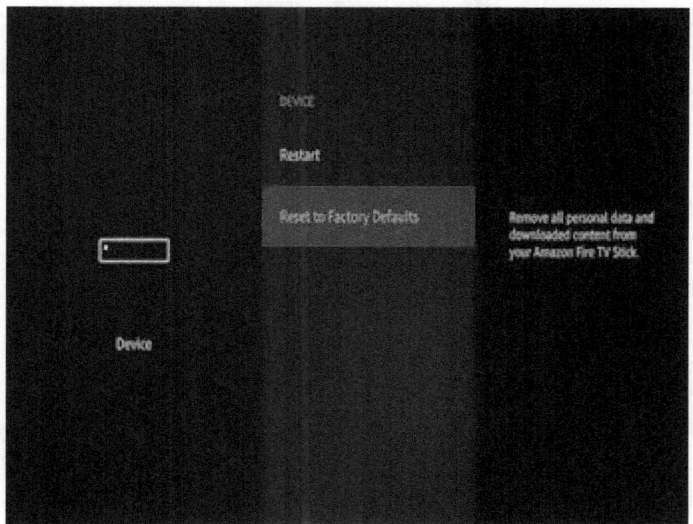

1. Go to the Settings, select Device, and choose Reset to Factory Default

2. Choose Reset

Your device will begin the process of resetting. The process can take up to 30 minutes. When the process is over, a prompt similar to the one at the beginning tells you to connect your remote.

CHAPTER 6

What to Watch with the Amazon Fire Stick

Now that we have explored the basics of the Fire Stick and how to get everything set up, it is time to understand what you are able to watch with Fire TV. Amazon is one of the power players in the digital media market due to the fact that they have a lot of games, music, TV shows and movies available for their Fire Stick. You have two options when it comes to accessing all the content Amazon has to offer; you can either purchase each of the titles you want to watch individually, or you can sign up for their subscription service.

With the a la carte options, you pick out the items you want to purchase one at a time. If you don't watch a lot of movies or shows, this can be a good choice because you are only going to pay for what you need. But if you like to watch a lot of movies and shows, it can get pretty expensive quickly.

A better option is the subscription service. You can get an Amazon Prime account for about $10 a month; you will get free shipping, free movies and shows, free books, and other great benefits all at the same time. You will not be limited to the number of movies and shows so if you like to relax each night with a movie or a show, this will be the most cost-effective method for you.

Channel Subscriptions

In addition to using all the great services that come with Amazon Prime with all the movies, shows, music and more, you are able to subscribe to other channels and watch them with your Fire Stick. Remember that you will need to download all the channels' apps from the Amazon app store. For example, if you want to watch Netflix on the Fire Stick, you will need to go to Amazon and pick out the Netflix app.

Auite a few of these media subscriptions are free to download, such as Hulu Plus and Netflix, but for some, you will need to pay each month to get the music and shows you want. No matter which app you want, you will be able to download it from the Fire Stick, and it is pretty easy to get it all set up.

There are a ton of things to watch with your Fire Stick, and the library of channels is already pretty large. You will find that Amazon is continuously adding more channels, so if one of your favorites is not there, you are sure to see it before long.

Movies and Shows

- Amazon Instant Video—this one doesn't need an app to get started, and you won't pay a fee each month to subscribe. You just need to choose the title pay for it and watch it.
- Amazon Prime Instant Video—this one has a fee each year, but you get to stream as many movies and videos as you like from a library of 40,000 titles. The app is not required since the Stick is already tied to this store.
- Netflix—if you already have a Netflix account, you will be able to download the app and watch unlimited shows with this streaming service without all the advertisements.
- Hulu Plus—this site is another great one for television episodes, and with a small subscription fee and the right app in place, you can watch it with the Fire Stick.
- Showtime Anytime—this site is a good one for comedy, sports, movies, and any other original programming from Showtime. You will need a paid subscription with cable or a dish plan to get this to work.
- Crackle—just like Showtime Anytime and Netflix, Crackle is going to offer original

programming, TV shows and movies. It is owned by Sony Pictures, so most of the media will Sony products. The service is free, but be aware that it has commercials.

- YouTube—this is a good source for free videos and music. It is supported by ads, but it has a big selection to choose from.
- Vimeo—this alternative to YouTube is a good video sharing site. You will be able to watch the videos for free, but there is a subscription if you want to upload your videos or watch them without ads.
- HuffPost Live—with this one, you get to watch Internet streaming of the Huffington Post covering news and politics. It is ad-supported and free.
- Flixster—this free channel of social networking allow users to share reviews and trailers of their favorite movies. This is the parent site of Rotten Tomatoes, and the app will give you access to both.

Business programming

- Bloomberg TV—this is a good standby for the business media world. There is a paid subscription if you want to avoid ads as well as a free version with ads.

- TastyTrade—this trading channel shows investing tips, money management and more about the financial world.

Sports Fans

Watch ESPN—this is going to stream all the products from ESPN in a simulcast, including ESPN, ESPN2, ESPN3, ESPNU, ESPNEWS and so on.

NBA Game Time—this will include statistics, schedules and box scores for all NBA games, plus highlights from each game. With an upgrade, you can watch the games live.

Red Bull TV—this channel will allow you to watch short and full movies on sports topics, breaking news and interviews.

Music
Prime Music and Amazon Music Library

Of course, the first place to look for music is your Amazon account. If you have already purchased music from Amazon, this will be available with your account online through the Amazon Music Library. To see the titles available, you check out www.amazon.com/musiclibrary. Sometimes, some of the CD's you purchased in the past from Amazon will still be available thanks to the AutoRip feature that comes with Amazon.

If there are other songs you would like to add to your account, you will just need to purchase them from Amazon. You can also use Amazon Prime Music to get free music to listen to on a regular basis. This is included in your Prime membership, and you can scroll through using voice command or by texting the name of the songs you want. It is easy to see which options are available for free with this particular service.

So, if you already have music on your Amazon account, you will find it via the Music setting on your Fire Stick. You can then access the library and all the music you have stored there. You can navigate through the music and click Select for the music you want at the moment. Once the track has started, you can leave the music menu to use some of the other features that come with the Fire Stick at the same time. Just use the right buttons on your remote to rewind, fast forward, pause, stop, or playback the music.

Qello Concerts

Qello Concerts provide HD footage of the greatest concerts, both old and new, in all genres. It is a music video channel with a lot of great live concerts that are sure to please everyone. You will find a lot of categories including opera, Pink Floyd, Beatles, Jazz and more. It does emphasize contemporary music most of the time, but it also is tailored to a wide range of tastes, so you are sure to find a lot of great music. You will need to register for an account; but it is free, and you can do it from the Home screen of Fire TV. You can also upgrade your account to get away from commercials and get more access to concerts than before.

Vevo

Vevo is the best music video site online, with music videos from two out of three major record labels. It is compared to MTV with a free ad-supported platform, and you can choose from their library of more than 75,000 music videos or stream their music video channel. The search function is a bit hard to use since it won't let you search by category, but there is a lot of great music to choose from. The app for Vevo is currently free, but there are 30-second ad spots that show up before the videos start.

Pandora

If you have a Pandora account, whether free or subscription based, you can use the app on your Fire Stick to listen to music in many different genres.

Listening to the Radio

Not only are you able to use the Fire Stick to listen to music from the Amazon Music Library and watch music videos like those listed above, but you will also be able to use some of the internet radio stations that Amazon has partnered with. The interfaces for these stations are similar, so we are going to look at some of the pointers to follow when you want to turn the Fire Stick into your own personal radio station.

Right now, the Fire Stick has three channels that are music only, and they are going to work pretty much the same way on Fire TV. You will notice that the categories are on the left, and then suggestions for what you might like to watch are on the right. On the left side, you can choose the genre of music, and this will make changes in the suggestions on the right.

If you are using iHeart Radio or Pandora, you will notice that you get the option of building up custom song lists from the musical libraries on these channels; these are called custom stations. This is easiest when you use your laptop because you are able to use the keyboard and type what you would like. Let's take a look at how you can create your own playlists when using the radio stations through the Amazon Fire Stick.

Pandora

Pandora is an Internet radio project that allows you to listen to the radio stations of your choice once you register for a free account. You can do this from the Fire Stick or your personal computer. You will be given the option of making your song lists you public so you, in effect, you have your own radio station, or you can keep it private such that you are the only one able to use it.

It is also possible to browse through the selection of music available on this channel and listen to any songs you want, rather than creating a playlist. One of the things you will find useful about this app is that you can place it on a sleep timer with alarm clock features, so it will alert you at certain times or turn off when you would like. You can get this app for free, but be aware that there are some commercials with the free version. If you would like to avoid these commercials, you can upgrade to Pandora One and listen to all the music you want.

iHeart Radio

This media giant has over 800 US radio stations on the iHeart Radio channel for you to listen to on your Fire Stick. You are able to listen live on an individual basis, or you can compile the content you want and make your own music station. iHeart Radio is popular because you don't need to register for the account and there is no fee (or even a paid upgrade) to worry about.

On this station, you will notice video ads on the live stations, just as if you were listening to music on the radio in your car or somewhere else; but if you create custom music stations, they are free of commercials at this time. The iHeart Radio Fire Stick channel is also set up with an alarm clock and sleep timer just like you find with the Pandora service.

You can see that there are a lot of great offerings that come with the Fire Stick; and if you already have a subscription to some of them, you can add the app to the Fire Stick and they will work well together. Amazon is always adding more of these channels, so if you don't see some of your favorites right now, it is sure to be added later on.

TuneIn

The third service you can use when you want to listen to the radio on your Fire Stick is TuneIn. This allows you to enjoy live listening on more than 100,000 Internet radio stations, including C-SPAN, NPR, ESPN, CBS and a selection of international stations. You can go through and look at more than 2 million archived interviews, concerts and podcasts.

To use this particular service with the Fire Stick, you will need to register for an account. The interface on TuneIn is pretty easy to use, and it enables you to set up an account and enjoy free radio, but it does have some basic ad support. In addition, if you would like to record the stuff you hear on TuneIn for later, you can do this for a onetime fee of $3.99; you just have to remember that you aren't allowed to share the recording with others and can't use it to make money.

Listening to music is a great way to use your Fire Stick efficiently. You can listen to some of your favorite songs from your Amazon account, tune into your favorite radio stations or combine a few to make a new one, or even listen to music videos all in one place. The Fire Stick brings all of this together to make a great new musical center of your favorites.

The Amazon Cloud

At this point, it is time to talk about the Amazon Cloud. This great storage device comes with your Fire Stick and makes it easier to back up your device and to keep your videos and other things stored in one place. The Fire TV Stick accesses the Amazon Cloud at any time to retrieve personal videos, photos, games, apps, movies and anything else stored there.

To get on the Cloud, scroll down your Main Menu and then choose the type of media you would like to retrieve. The Fire Stick will be able to pull it down from the Cloud so you are able to watch as you wish.

You don't have to worry about setting up the Cloud system or purchasing more space right away. Whenever you open an account with

Amazon (something you already did when you ordered the Fire Stick), you are going to be given 5 GB of storage. You can place any type of media you want in the Cloud.

For those who go through a lot of files and like to download tons of shows and movies, you can have than the 5 GB of storage. There is a low annual fee for doing this, and the amount changes based on the amount of space you would like to purchase. One thing to note is when you purchase media from Amazon, it is going to be stored in the Cloud account for free, without taking up any of your 5 GB of quota. You can also delete files you no longer want or watch and free up some space to stay within the free 5 GB for as long as possible.

There is so much to do when it comes to working with the Amazon Fire Stick. Many channels are already in the apps with Amazon, and you are able to pick and choose which ones are best for you. Some choose to just stick with the Amazon free services and watch their great shows and movies, while others may have a Hulu, Netflix, or other account to hook up to the Fire Stick so they can watch all in one place. The choice is up to you, and with a growing library of new apps and channels added to the platform all the time, it won't take long before all your favorite channels are available for this device.

CHAPTER 7

Streaming Tips for the Fire Stick

Your Fire Stick has so many cool features. In this chapter, you will find interesting other features of your device. You will learn ways of eliminating ads.

Change the name of your device

Every device has its own name and account. If you have many devices, you may get confused with the names, but you can change the name of your devices. You need to go to the browser. In your browser, go to Content and Devices Management >Your devices> Edit. Change the old name to the name you want.

Deleting Voice Recordings

You may not like Amazon storing your voice recordings. However, this lets you improve accuracy when you listen to your slow speech. Even if it "can worsen your impression of using voice functions," voice recordings can be deleted. For people concerned about privacy, this option is very convenient.

To do this, go to the browser. Manage Your Content and Devices > your device (Fire Stick)>Voice Record Management> delete.

You can use Alexa to delete voice searches one at a time. Open the Amazon Alexa application>Settings>History> entry>Delete.

Talk to your TV

Just a couple of years ago, it would have sounded strange. But with new technology, you can do it. If you have Amazon Echo connected to your Fire Stick or have a voice assistant voice remote, you are able to talk to Alexa. It will find a video, turn on an app, and even order food.

Disabling auto play video

When you press pause and do not perform any action for a long time, the automatic playback will start. This will include not only video but also audio.

You should select "Settings" and then "Preferences," find "Featured Content" there and disable the option, "Allow Auto start Video." I advise you to also disable Allow Audio Auto play.

Scrutinizing What You are Watching

If you want to view information about a movie or actor, this is available using the X-Ray tool. Press the "up" and "down" buttons, and scroll through the information. You can find out about the music being played on any scene and receive trivia as you continue to watch.

Stop Advertising in its Tracks

Unfortunately, your device does not block ads. It only reduces the amount of advertising.

However, Amazon has now made ads more private on your Fire device. With the new software, advertising can be partially gotten rid of. You can remove ads based on specific interests or those based on what you see in different applications. To do this, you need a software version not lower than 5.2.1.1. Your views allow advertisers to show you what you have been looking for.

In the settings, find the item "Interest-based Ads" and click it off. This will protect you from a lot of advertising but will not switch it off completely. After that, your ads will be more random, and there will be less data. If you like privacy, this is a great option for your device.

You cannot refuse advertising, but you can prohibit the device from tracking you for promotional purposes. Turn off targeting by selecting Settings. There you should find Preferences. Then select Advertising ID and switch Internet-Based Ads to Off. In addition, turn off Internet advertising.

Applications for blocking ads

Blokada

Officially, you can't disable ads on your device. But you can use applications to get rid of advertising by installing Blokada.

Amazon does not offer any direct ad blocking tools, but you can download various apps to block ads on your TV. Let us take a look at how to block ads on your Amazon Fire ,using one of the popular applications called Blokada.

You have to download and install the application. To get rid of advertising, start by installing the application. In order to install Blokada on your device you need to do the following:

On the main screen of your Amazon Fire, click "Settings." This will open the the Settings for your device. In Settings, click "Device."

Clicking on "Device" will take you to a new list, which will show detailed information regarding your device, as well as allow you to change some of the settings.

Then you need to find and select "Developer Options." You need to enable "ADB Debugging" and "Apps from Unknown Sources."

Do not forget that in the list of devices, you must select "Developer Options."

You need to enable ADB Debugging and "Apps from Unknown Sources." This will help download Blokada and use it.

Then you download the application, "Downloader." If you do not have this application, then download it. To do this, go to the Home screen, search for the Downloader, download and install it. Open the downloader. There enter the URL for Blokada and click Go.

If you did everything right, you will see a page for downloading. After downloading this application, you need to install it on your device.

After installing Blokada, on the welcome screen check the "I'm new" box. This great app will help get rid of ads.

After you've download Blokada, quickly install it and on the welcome screen, switch the "I'm new" button and click "Continue." This will open advanced lock options. In the application menu, enable all features except "Notification." You do not need it. If you enable notifications, you will receive them every time an advertisement is blocked.

Next, you need to find the "Power" button in the lower right corner. This will enable your application.

After you do everything, you will get a view of your content. This application saves you from advertising.

Stopad

This is another popular application. When we decided to purchase this device, we expected to receive free viewing. But advertising has already penetrated this device.

Attention! The application you are offered is not official for your device!

First, you need to allow your device to accept third-party applications from any source. The installation principle of this application is the same as the one described above. Go to "Settings," then go to "Device," and then "Developer Options." This will lead you to new settings. Turn on "Apps from Unknown Sources." Next, to download you need the program "Downloader." If you do not have this, then install it.

After that, open the search in the "Downloader" program, enter "Stopad" in the URL Search: http: //stopad.io in the search for the URL of the downloader and click "Go."

Download the program to your device. Adopt and open the app. You need to click on "Enable Stopad." Your device is now free from advertising.

CHAPTER 8

Kodi and Fire TV Stick

The Amazon Fire TV Stick is one of the best ways to make the most of your Kodi experience with your HD television. The installation of Kodi on your Fire TV device is quick and easy. Kodi is highly compatible with Amazon Fire and can be fully installed in only a few minutes with just a few fairly modest steps. This can be done by many methods, one of which being the side load method.

Even though most of the methods can be done relatively quickly, side loading is the most popular method of installation. Side loading is said to be similar to downloading or uploading something. It is seen as a shortcut to the conventional method generally used to install an application. Using the side loading method does not require any rooting of the device or major modifications. It also does not require hacking in order to install Kodi on the Fire Stick TV. This method can be used for the Fire Stick TV and the Fire Stick.

The Kodi Installation does not require any advanced knowledge. Its simplicity and ease of installation are some of the reasons why Kodi is so popular today. Many of the methods are based on side loading, which does not require additional or advanced hardware.

Kodi can be installed on Amazon Fire TV using apps available in Google Play; these same apps are made available for download on Kodi's website, which is Kodi.tv. Below we will discuss several methods to install Kodi on your Amazon Fire TV in just minutes.

Manual methods of installation are slightly more challenging than using apps, but they yield the same results. If you are computer savvy, you will have Kodi installed on your Amazon Fire TV Stick within just a few moments when using the manual method.

Installing Kodi On Your Fire TV Stick

There are different ways of installing Kodi on your Fire Stick TV Device. In case you are not using a stick, but instead are using Fire TV, it does not require rooting. It simply uses Kodi for Android.

How to Install Kodi using Downloader on Amazon TV

To begin installing Kodi using the Downloader on Amazon TV, plug your Amazon Fire TV Stick into the television if this has not already been done. After plugging your device into the television, if you have not already done so, you will be prompted to sign into Amazon.

Next, you will have to go through the Settings Menu. From the Settings menu, scroll to the right of your screen and select System from the list of options. From the designated System menu, click "Developer Options." You will then turn on the following options within the developer options.

ADB Debugging: ON

USB Debugging: ON

Select OK to get rid of the warning message that surfaces on your screen.

Navigate out of the Settings menu and scroll to the Search area of your Amazon device.

Once you have completed the above steps, search for the application named "Downloader." Select the desired application. This Downloader app can be found in the Apps & Games Category.

When the Downloader app opens, key in this URL: "http://bit.ly/kodi181arm" to download and install Kodi on FireStick.

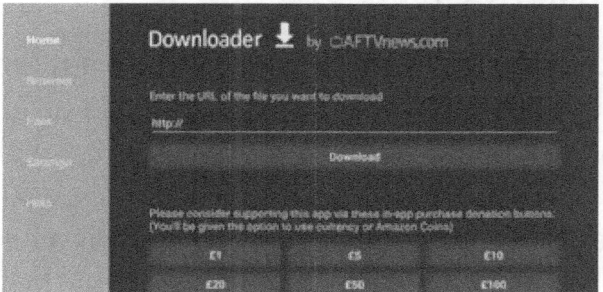

Then select "DOWNLOAD"

Wait until the downloading is over. In case the Kodi installer does not start, select "Open App."

When it launches, choose "Install."

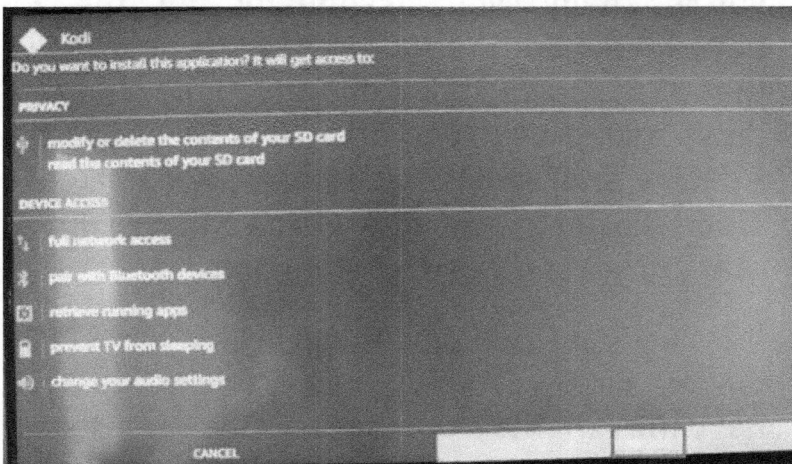

Hold on for a little more until it is installed. Select "Open App."

From here you will install Addons for the TV to protect Kodi streams.

Installing Kodi with a USB

The cool thing about Kodi is, if you're not familiar with the different ways to do it or side load it, it's important to consider the USB option. You can put the Fire Stick into a USB port, and it will come up as storage. You can't go anywhere on the directory, but you can see some of the items of the customer.

From here, you can side load it directly from the cloud. You may realize that some of these methods are incompatible with the app store, but you can get some there directly there. If you're ready, you essentially put the ES downloader and the explorer to a device, and on that USB, you can put it directly to the cloud for download. Go to apps and games, choose the Kodi file, and from there, upload it directly to the cloud. It's pretty amazing what it can do, with so many great and worthwhile actions.

If you want to download it via the APK downloader browser extension, make sure you have the APK file from the internet. It's a decent option, but here's the problem: sometimes these extensions may not work. While it's downloadable, it may not work via the app's UI.

You can always get it through the install with a USB port as well. If you go to the website and search, you'll see it. The best one for this is the Direct APK downloader because it actually can download away from the internet directly to the android devices; and if you already have Google play, it will use the credentials to help with this.

At this point, once you've got the PK fully downloaded, you literally just copy and paste it directly to the Fire Stick. You can do this as a download or in the download folder. Then you want to open the apps in the same way you would with ES file explorer, or even downloader, and from there, work to install the APK. You literally open up downloader, choose Kodi, install and launch it, and there you go!

USB side loading is probably the best way to do it if you're going to be using it for developer reasons, or if you don't want to deal with the hassle of side loading various apps with small URLS that might not work. USB storage devices work similarly to Fire Stick devices. There are some devices you connect to, and once the USB device is in there, you can boot the PC. The USB media devices are good for storing media, but the Fire Stick, due to its size, may not be ideal for this. However, it can be used with USB, and also connected to the HDMI, so you can stream media. It's the perfect way to install Kodi.

What Kodi is

Kodi is a free and open source piece of software designed specifically for home entertainment, with the fans of television content, films and movies in mind. Kodi was first developed in the year 2003 and was originally named the Xbox Media Center (XBMC), as it was mainly developed to use the Microsoft Xbox gaming console as its platform. Over the years, the software evolved and expanded to other platforms as well and was rebranded and launched in its current incarnation as Kodi in 2014.

It can be installed on almost any multimedia platform, from phones and televisions to tablets, and can even use streaming sticks as a platform. Kodi is not limited to streaming but is also used to create what is known as a home streaming hub, allowing multiple devices on a single network to become connected to share a common video, music and picture library. Kodi does not produce original content itself, but it allows users to transfer content from their own storage media or platforms, so they can transfer the contents of their DVDs, Blu-ray discs or their USB drives.

Kodi is also capable of turning into a type of planner, capable of recording content for later viewing. Plus, it can organize your media into easy-to- access forms. In addition, Kodi's functionality can be enhanced by multiple third-party plug-ins and add-ons that are available. These plug-ins and add-ons add various capabilities such as access to YouTube, Vimeo and even Netflix content.

Kodi is also a media player with access to various codecs, meaning that it can play almost every format of video and audio, removing compatibility issues and allowing almost all content to be played in a single location.

Why use Kodi

Kodi offers its users freedom, flexibility and convenience in setting up their home entertainment systems. The sheer number of features that Kodi makes available to users makes it one of, if not the most popular, home streaming and multimedia service software packages online. The fact that it is a free and open source contributes to its popularity, as well as the fact that it is capable of being installed on almost all forms of multimedia devices, from televisions and laptops to smartphones.

Many use Kodi to develop an integrated multimedia hub with access to content from every device on its network. However, some users use Kodi in order to access content from the internet. While Kodi itself does not provide any content, third-party plug-ins are compatible with Kodi that allow users to access content from the internet, such as from Netflix, Hulu, and HBO Go. This software is usually made available for free, allowing the end user to enjoy content that is normally locked behind a paywall; and this is something that makes Kodi popular with certain users. However, if someone wishes to use the Kodi software on their Amazon Fire Stick, they cannot use it through a straightforward download, as it is not available in the Amazon application store. The following will provide some methods of installing Kodi on your Amazon Fire Stick.

Methods of Installing Kodi

Using a VPN

There are multiple VPN clients available online for free. One of the more popular VPNs is IPVanish, but most VPNs available online provide for the basic needs of a user, primarily safety and privacy. Most VPN clients only require a quick download and will set up automatically upon use, needing little to no technical knowledge. Certain commercially-available routers come pre-installed, but these often require more technical know-how to set up, as a small mistake may leave the network vulnerable to online attacks.

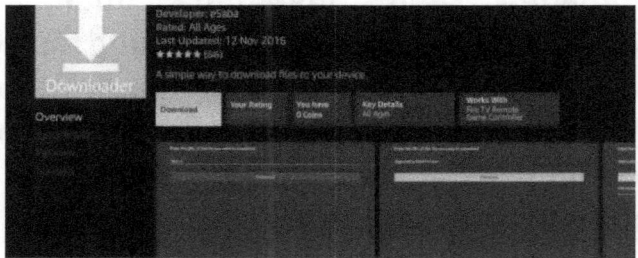

Using Downloader

The first step when installing Kodi is to make sure the device is turned on and connected to the internet. The next thing is to go to the settings tab, then within it, the device tab. Once you have navigated to this point, select developer options (in some devices Debug Options), and allow the apps from Unknown Sources to be turned ON to allow third-party applications to be installed. If you are worried about the possible safety issues, at this point, a pop-up should appear warning you about unknown sources. You can always navigate back after installing Kodi to turn off the permission, but we need it set to ON for now.

Return to the Home screen and navigate to the Apps portion, and through the sub-menu that appears, select the Categories portion. Select Utility, and then search for the Downloader application. Select it and press "Get" to allow your Fire Stick to download and install this application. Once the application has been downloaded and installed, open said application and a popup screen asking for a URL should appear. The most up-to-date version of Kodi is currently Kodi 17.5, released October 24, 2017; but it still has issues, so we advise downloading the stable version, Kodi 17.4.

In order to do this, type "bit.ly/174kodi" or "bit.ly/kodi17apk" into the URL bar to allow Downloader to download Kodi Krypton 17.4. The version of Kodi you selected should begin downloading. However, there may be a pop-up notice informing you that JavaScript is not enabled. In this case, navigate to the settings sub-menu and find the option to enable JavaScript there. Once finished, select install and wait for the program to unpack and install on your Fire Stick. When it has finished installation, select Open to ensure that Kodi is installed properly. The first time the program opens and initializes may take some time, so don't worry if it takes more time. After waiting, the Kodi Home screen should appear. At this point, Kodi has been installed and you can begin using the software.

Using ES explorer

The first step when installing Kodi is to make sure that the device is turned on and connected to the internet. The next step is to look for the settings tab, then within that, the device tab. Once you have navigated here, select developer options (in some devices Debug Options), and then allow the Apps from Unknown Sources to be turned ON, to allow third-party applications to be installed. If you are worried about possible safety issues, a pop-up should appear, warning you about unknown sources. You can always navigate back here after installing Kodi to turn off the permission, but we need it set to ON for now.

Return to the Home screen and navigate to the Apps portion, and through the sub-menu that will appear, select the Categories portion within the sub-menu. Select Utility, and then search for the ES Explorer application. Select it and press "Get" to allow your Fire Stick to download and install this application. Once the application has downloaded and installed, open said application and a popup screen asking for a URL should appear. The most up-to-date version of Kodi is currently Kodi 17.5, released on October 24, 2017; but it still has issues, so we advise downloading the stable version, Kodi 17.4. To do this, type "https://troypoint.com/kodistable" or "bit.ly/kodi17apk" into the URL bar to allow the ES Explorer program to download Kodi Krypton 17.4. The version of Kodi you selected should begin downloading. However, there may be a pop-up notice informing you that JavaScript is not enabled. In this case, navigate to the settings sub-menu and find the option to enable JavaScript there. When finished, select install and wait for the program to unpack and install on your Fire Stick. When its done installating, select Open to ensure that Kodi installed properly. The first time the program opens and initializes may take some time, so don't be concerned if it takes more time. After waiting, the Kodi Home screen should appear. At this point, Kodi has been installed and you can begin using the software.

CHAPTER 9

Troubleshooting Options

It is rare for users to have any set-up issues with their Fire Sticks. Amazon has gone to great lengths to make the devices as user-friendly as possible, and this really shows with the interface on their Fire operating system. The most common issues people have with the Fire Stick stem from either Wi-Fi connectivity or peripheral devices, like remotes and controllers.

The "Help" section in the settings interface is a good first place to look if you are having problems with a specific aspect of your Fire Stick's operation. You can see step-by-step instructions for fixing common problems, designed to be easy to understand and follow.

While it's become something of an IT cliché, there is some wisdom to the idea that power cycling a device can solve its issues. Especially if you use a lot of different apps, data can accumulate inside your Fire Stick's memory as time goes on, and this can result in slow performance or other operation problems. Rebooting the device clears out this backlog, letting you start fresh. It's a good first step to take any time you're having problems.

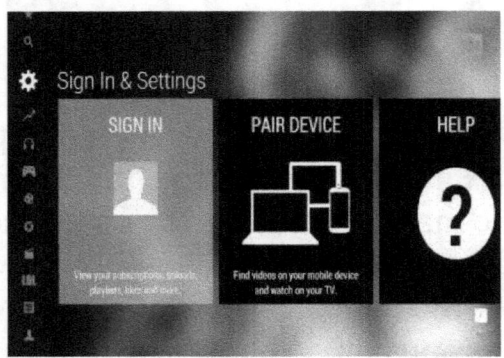

You can restart your Fire Stick by going into the settings, then selecting "System" followed by "Restart." Alternatively, you can do it with the remote by holding down the play/pause and select buttons simultaneously. After about five seconds, the screen should go black, after which you'll see the Fire TV logo as the system re-boots.

If your system seizes up and won't re-start by either of the above methods, you can do a hard reboot by unplugging the power source. Remove the power cord from the back of the Fire Stick first, then pull it out of your TV. After a short time, plug it in.

One common issue with the Fire Stick is the problem of keeping the remote or controller paired with the device. The good news is, this is very easy to fix. Simply hold down the Home button on the remote for between five and ten seconds, until the device is recognized by your Fire Stick. You can also use this method to connect a new remote or switch between controllers when you want to switch to a game.

Wi-Fi Fixes

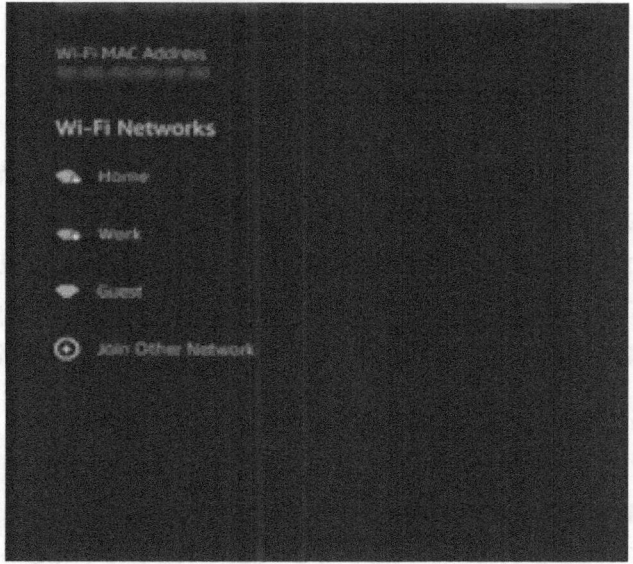

Streaming issues, like buffering, are almost never a result of the Fire Stick itself, or even of the app you're using to view content. The majority of times, there are problems with streaming content it is because of an issue with the Wi-Fi network.

Wi-Fi connectivity issues can bring up a variety of different messages on your Fire Stick interface. While the messages can help you diagnose the root of the problem, your first step when having issues with connectivity should be to determine if they stem from the network itself or if they are specific to your Fire Stick. Try using the internet on the network the Fire Stick uses with another device in the house. If it works, troubleshoot the Fire Stick; if it doesn't, start with your router.

If you get the message "No Connection" or "Not Connected to the Internet," this is almost always an indication that the Fire Stick has found your wireless network but isn't able to access the internet through it. Restart your modem and router to see if this is the problem.

If you get one of these messages and the other devices in your home are accessing the internet without an issue, the problem may be the distance between your router and the Fire Stick. Try rearranging things so that the two are closer together, and see if this alleviates the problem. Also make sure your router isn't close to anything that can block the signal, like thick walls or large pieces of metal.

Where your router is in relation to other devices can also have an impact. If your router is currently surrounded by things that use the internet, shift some of those objects to another room, farther from the router, and you may see an increase in the signal strength. Putting the router higher in the room than other devices may also help. A flat, high surface like the top of a bookshelf is the ideal location for a router.

Upgrading to a higher internet speed will sometimes solve connectivity issues, but if you have an especially large home, you may find the reach of the signal is more of an issue than bandwidth. You can buy Wi-Fi extenders or additional routers to help the signal better reach all of your devices.

When the message comes up, "Check your network status," this could be a problem with either your internet connection or your Fire Stick device. You can get more details about the nature of the problem by selecting the option on the message to open the network settings. Use the play/pause button to launch the program, "Network Status Tool." If you've already closed the message or want to do this pre-emptively, you can find this in your settings under the "Network" sub-menu.

The Network Status Tool checks two things: that your device is able to obtain a connection to the network and that the network itself has a connection to the internet. Once it has diagnosed the origin of the issue, it will recommend a series of solutions for you to follow to correct it.

A low-quality signal can be almost as bad as no signal at all. If you're usually able to get a good signal on your Fire Stick but suddenly are having issues with buffering or dropped feeds, first double-check that the power cord is fully connected. If your Fire Stick is trying to run off of the power it gets from the TV, it may seem like it's functioning but not have enough juice to operate correctly.

Users with multiple TVs in their home may notice that the Fire Stick seems to work better on some than others. This can be a result of where the TV is located in relation to the router, but it can also be caused by the TV itself. Certain brands and designs partially block the Wi-Fi signal from getting to the Fire Stick. You can use the HDMI extender that comes with the Fire Stick to move it farther away from the TV and see if this helps alleviate the problem.

You should also consider whether you've added any new devices to your network recently. Think of your Wi-Fi signal as a stream of water. It has a lot of power when flowing into a single source, but the more directions you divert it to, the smaller the amount of water will be flowing into each channel.

Reducing the number of directions your Wi-Fi signal must go can help increase the strength of the signal to each individual device. If there are old devices in your home that you no longer use (or use rarely), remove them from your Wi-Fi network. You can do this through the device itself, either by putting it into "Airplane mode" or disabling the Wi-Fi connection.

The kind of antenna your router users will also have an impact on how the signal gets to your devices. Most routers come with an all-purpose omnidirectional antenna installed. This is the best choice for most users, but if you want to, instead, focus the entire signal to one specific device, you can buy a single-directional antenna. This will double the signal strength over an omnidirectional antenna and send it all to one source.

You should also make sure that you're the only one using your Wi-Fi. If you live in an area close to other people, make sure you set password protection on your network. Not only will this prevent freeloaders from mooching off of your bandwidth, but it will also make sure your personal information isn't accessible to others.

If nothing else seems to work and you don't want to upgrade your wireless plan, you can rig up a DIY wireless enhancer by positioning a piece of metal behind your router or around your antenna. While it might look a bit kooky, this kind of system has been field tested and proven effective. You can find a variety of tutorials online that use everything from pieces of aluminum foil to flattened out beer cans to enhance the strength of their wireless signal.

Conclusion

If you love watching television shows and movies and would like to watch them all on your TV, the Amazon Fire Stick is one of the best products to get this done. It is simple to use, has a lot of options, and it comes with all the high benefits you have come to expect from Amazon technical products.

This guidebook spent some time talking about how to use the Fire Stick; and even if you don't have any technical skills to start with, you will find it is pretty easy to set up the Fire Stick and watch some of your favorite television shows and movies, while also enjoying games, apps, photos, music, and so much more. You simply need to plug in the Fire Stick, pair it up with the remote or controller you want to use, and then purchase the right apps to watch the channel featuring your favorite programs. You can add things to your Amazon Cloud service, and you will find that all your media needs will be in one place.

When you are ready to watch movies, shows, games, music, and more in just one place, get the Amazon Fire Stick. It is one of the best devices on the market for what it can do, and you will appreciate how easy it makes your life. Take another pass through this guidebook and review everything you need to know to get started with the Amazon Fire Stick today.

One last thing

If you enjoyed reading this book I would kindly like to ask you to leave a review on Amazon. It'd be greatly apprechiated.

www.ingramcontent.com/pod-product-compliance
Lightning Source LLC
Chambersburg PA
CBHW072214170526
45158CB00002BA/599